The Blue Helix
Guide to Scientific Experiments

The Blue Helix
Charleston, SC

THE BLUE HELIX

Copyright © 2012 Jeffrey Scott Coker. Published by The Blue Helix. All rights reserved. No part of this publication may be reproduced, transmitted, digitized, or recorded in any form without written permission from the author and publisher. This publication is for informational purposes. The author and publisher take no liability for the use of this information and any claims that arise there from. Photo credits (©iStockPhoto.com): Girl with magnifier - Izabela Habur; Molecule - Alengo; Lab - Nicholas; Thinker - Dan Wilton; Laptop girl - Justin Horrocks; Die - Onur Döngel; Circuit girl - Nullplus; Cycle - ewg3D; Equipment - Alexey Dudoladov.

ISBN: 978-1-937109-02-8

Contents

Introduction to Experimentation	1
Step 1 - Come up with a Topic	5
Step 2 - Ask a Scientific Question	11
Step 3 - Develop a Hypothesis	12
Step 4 - Design an Experiment	14
Step 5 - Observe and Record Data	17
Step 6 - Analyze Data	19
Step 7 - Make Conclusions	25
Step 8 - Present Results	27
The True Scientific Method is a Cycle	30
Guidelines for Achieving Great Experiments	31
Scientific Notebook	32
Lab Report Template	84
Units of Scientific Measurement	90
Glossary	92

Introduction to Experimentation

Imagine being the very first person on Earth to understand some particular part of nature. You alone can see this small bit of knowledge and how it relates to everything else. It is like having your own private window on the world. Even when it involves something small and simple, discovery is a wonderful and exciting feeling!

Despite how scientists are often depicted, discoveries are not usually made by eccentric geniuses. Instead, discovery is a very attainable process for everyone. If you ask focused questions and pursue answers in a systematic way, then you quickly surpass the limits of human knowledge and enter the realm of true discovery.

The process of discovery involves doing *experiments*. A scientific experiment involves **systematically changing something and then observing the effects of the change**. Sometimes experiments are conducted in laboratories in very controlled situations, and sometimes they take place in situations that are harder to control (i.e. ecological experiments). Experiments expose elements of nature that would not be apparent otherwise. They are a powerful way of making discoveries and generating new knowledge.

What is an experiment?

> A situation where something is systematically changed and then the effects of the change are observed.

Understanding experimentation is crucial for scientists and non-scientists alike. For example, advertisers constantly bombard us with claims about their products. If you understand what makes a good experiment, you can make better decisions in many areas of your life, including personal health, diet, finances, and politics.

Consider the following "Guidelines for Achieving Great Experiments":

1. Ask a very specific, testable question.
2. Test a control for comparison (a group that does not receive the experimental treatment).
3. Use a sample size large enough to allow firm conclusions.
4. To understand a whole population, obtain a random sample of that population to avoid bias.
5. Replicate each part of the experiment (at least 3 times).

Imagine that you want to understand the effects of high nitrogen levels on tadpole growth. You plan to put tadpoles in three bowls of water with high nitrogen to see what happens. What is missing from this experiment?

GUIDE TO SCIENTIFIC EXPERIMENTS

> A control group. Use six bowls and add nitrogen to three.

More "Guidelines for Achieving Great Experiments" include the following:

6. Hold all variables constant between trials except the variable being tested.
7. Collect quantitative data whenever possible.
8. Measure using metric units.
9. Gather data carefully and accurately.
10. Be objective and honest.

Imagine that you compare bacterial growth in the presence of four different antibiotics by recording which antibiotic allows the least growth. What could be improved in this experiment?

> Make quantitative measurements of the growth.

The following sections will take you through the scientific process, from coming up with a topic through presenting results. As you move through this process with your own experiment, there are three things that you should do repeatedly.

- Find lots of background information to help you. If you take the time to learn about the topic and previous experiments that have been done, then you can make your own experiment much better. Ideally, you want your experiment to build on, and not just repeat, what others have done. Remember to record where all of your information comes from (author, title, publisher, publication date, website, etc.).

- Get constant feedback from others and talk about your work with your peers and a mentor. Modern science requires collaboration and cooperation... it is very social!

- Be creative. The text that follows is written in "steps" to help you organize yourself, but you should not feel overly confined by the order of the steps. Real scientists do not usually work that way. Instead, scientists have to be highly creative and adapt their experiments as they understand more and more about the unknown.

GUIDE TO SCIENTIFIC EXPERIMENTS

Step 1 - Come up with a Topic

Coming up with an idea for an experiment is easy... just S.I.T. and think! S.I.T. stands for the three common categories for how scientists choose topics for their experiments: subjects, issues, and tools. As you move through the rest of this section, write down any ideas you have for possible experiment topics.

Subjects

Is there a particular subject that you're interested in? Every subject you could possibly think of contains many worthwhile experiments. For example, someone interested in guitars might design experiments involving how different types of strings affect sound, the effectiveness of computer programs for teaching guitar, or the mathematics underlying popular and unpopular music. A list of subjects is provided below to help you brainstorm possible experiments.

Animals	Electricity	Learning	Space
Behavior	Exercise	Math	Sports
Computers	Food	Medicine	Technology
Cosmetics	Flight	Music	Video games
Crime	Forests	Plants	Water
Disease	Lasers	Pollution	Weather

Issues

Is there a particular issue that interests you? Poverty? Health? Child safety? Education? The environment? Many scientists design experiments that will help solve global problems. For example, someone interested in hunger issues might design experiments involving the growth of corn in drought conditions, the effectiveness of different methods for transporting grain, or the dietary choices of people in different socioeconomic classes. A list of global issues is provided below to help you brainstorm possible experiments.

AIDS	Crime	Energy	Population
Aging	Culture	Environment	Poverty
Agriculture	Decolonization	Health	Security
Atomic energy	Drugs	Hunger	Sustainability
Child safety	Economy	Malaria	Terrorism
Climate change	Education	Peace	Water

Tools

Is there a particular tool or measuring device that you would like to use to gather data? Many scientists become experts at using certain tools and then come up with experiments to do using those tools. For example, someone with access to microscopes might investigate the number of guard cells on different types of leaves, the types of microorganisms present in different bodies of water, or the geometric patterns of different natural fibers. A list of inexpensive measuring devices is provided below to help you brainstorm possible experiments.

THE BLUE HELIX

Measuring Tool	Measurement	Where to Find
Albustix	Protein in urine	Pharmacy
Camera	Records images	General store
Digital multimeter	DC/AC voltage, etc.	Hardware store
Galvanometer	Electrical flow	Hardware store
Glucose monitor	Glucose in urine	Pharmacy
Graduated cylinder	Liquid volume	Science class
Home drug test	Drugs in body	Pharmacy
Litmus paper	pH	Science class
Peak flow meter	Lung capacity	Pharmacy
Pedometer	Distance traveled	General store
Pressure gauge	Air pressure	Hardware store
Protractor	Angles	Office supply
Rain gauge	Amount of rain	Hardware store
Ruler / tape measure	Length	Office supply
Scale / balance	Weight / mass	General store
Sieve	Particle size	Hardware store
Speedometer	Speed	Specialty store
Thermometer	Temperature	Hardware store
Water purity test	Chemicals levels	Hardware store

GUIDE TO SCIENTIFIC EXPERIMENTS

The measuring devices below are more expensive (over $50).

Measuring Tool	Measurement	Where to Find
Anemometer	Wind speed	Specialty store
Barometer	Air pressure	Specialty store
Impedance test	Body fat	Specialty store
Cholesterol monitor	Cholesterol	Pharmacy
Dosimeter	Ionizing radiation	Specialty store
Voltage gauge	Electrical voltage	Hardware store
Spectrometer	Properties of light	Specialty store
Hygrometer	Humidity	Hardware store
Incident light meter	Light on a surface	Specialty store
Reflected light meter	Light reflected	Specialty store
Refractometer	Salinity	Specialty store
Sphygmomanometer	Blood pressure	Pharmacy
Microscope	Tiny objects	Science class
Multistix	Chemicals in urine	Pharmacy
Telescope	Celestial objects	Specialty store
Tensiometer	Soil moisture	Specialty store
UV light meter	UV light	Specialty store
Wood moisture meter	Wood moisture	Hardware store

The lists above demonstrate just a few of the possible topics for an experiment. There are lots of other possibilities. Once you S.I.T. and think, you will find that coming up with a topic for worthwhile experiments is easy!

GUIDE TO SCIENTIFIC EXPERIMENTS

Step 2 - Ask a Scientific Question

Once you have selected a topic, the next step is to ask a good question. Ask something sophisticated that you are genuinely interested in answering.

The key to asking a scientific question is specificity. Your question should be specific enough that it suggests how you will collect data. Rank the following questions from best to worst.

  A. Do plants grow better in light or dark?
  B. Do plants grow taller in light or dark?
  C. Do corn plants grow taller in two weeks of constant light or two weeks of constant dark?

> C, B, A. Option C is specific enough to lead to an effective experiment. Option A is too vague, since "better" could mean many things. Option B is better than A, but still doesn't include important details.

There are lots of different plants and lots of ways to measure "growth," so being specific is essential!

Step 3 - Develop a Hypothesis

A hypothesis is a statement of what you think will happen in your experiment. It doesn't matter if you end up being correct or not. The purpose of making a hypothesis is to inform your experimental design. In other words, your experiment should be designed to clearly show whether your hypothesis is correct or not. Based on the questions in the last step, possible hypotheses include the following:

    A. Plants in the light will grow better.
    B. Plants in the light will grow taller.
    C. Corn plants will grow taller after two weeks of constant light than they will after two weeks of constant dark.

Which of these is the best hypothesis?

> C. Option C is clear and specific. A good experiment will probably be able to show whether or not this hypothesis is correct.

As you can see, starting with a good scientific question leads you toward a more sophisticated hypothesis and eventually a much better experiment.

Step 4 - Design an Experiment

When designing an experiment, your goal is simple - test your hypothesis. This simple goal sometimes can be achieved with a simple experiment. Other times, it takes a more complicated experiment. You want your experiment to be as simple as it can be while still providing a definitive test of your hypothesis.

In most experiments, you will want to create a situation where only one thing changes (the *independent variable*) so that you can observe its effects on something else (the *dependent variable*). When only one thing changes, you can be more certain that it is causing whatever effects you observe. However, only changing one variable in some experiments (i.e. those involving ecology, education, or behavior) is often very difficult and sometimes impossible. In these cases, you just have to do the best you can.

As you set up your experiment, remember these "Guidelines for Achieving Great Experiments" that were mentioned earlier:

GUIDE TO SCIENTIFIC EXPERIMENTS

- Test a *control* for comparison (a group that does not receive the experimental treatment).
- Use a sample size large enough to allow firm conclusions.
- To understand a whole population, obtain a random sample of that population to avoid bias.
- Replicate each part of the experiment (at least 3 times).
- Hold all variables constant between trials except the variable that is being tested.
- Collect quantitative data whenever possible.

Which example below would be most appropriate as a preliminary experiment?

    A. Grow 2 corn plants in the dark and 2 in the light.
    B. Grow 5 corn plants in the light and 5 in the dark. Repeat the experiment 3 times.
    C. Grow 10,000 corn plants in the dark and 10,000 in the light. Repeat 5 times.

> B. The sample size in A is too low and there are no replicates. Experiment C wastes time and materials. Experiment B is a reasonable way to begin answering a question.

Let's return to the corn plant experiment. There are lots of questions that need to be addressed. How many plants should be grown in the light and in the dark? How many plants should there be per pot? How big should the pots be? Should the pots have holes in the bottom? How many pots should there be per tray? How often should they be watered? What type of soil should be used? The best experimenters take the time to brainstorm and troubleshoot such questions, since they can completely change the outcome of an experiment. It is common to do preliminary experiments before the main experiment is done to help sort out some of these questions. This aspect of experimental design is highly creative, and often separates mediocre experiments from brilliant ones.

Step 5 - Observe and Record Data

The whole purpose for doing an experiment is to discover the truth about something. If you do not carefully, accurately, and honestly collect data, then you are just wasting your time. Good experimental scientists make accurate observations and write (or type) them down immediately. Otherwise, your memory of the observations will fade before your record them.

When should you write down observations?

> As soon as you make them.

Obviously, you want to record observations related to the hypothesis you are testing. For example, in the plant growth experiment mentioned above you would measure height. However, you should never feel confined by a particular hypothesis. The best scientists are open to observing the unexpected. The reality of experimentation is that unexpected results are normal. You think that either A or B will happen, and then you find that C happens (and then you have to do another experiment to figure out what C really means).

Keep in mind that there is absolutely nothing wrong with an experiment that doesn't turn out the way you expect. To a scientist, it doesn't matter if a hypothesis ends up being correct or incorrect. In fact, most really interesting experiments have unexpected or unpredictable results. What matters is that the experiment sheds light on the original question. The only bad experiment is the one that doesn't teach you anything.

Is your experiment better because your hypothesis turns out to be correct?

> No. What is important is that you learn from the experiment.

Lab notebooks

Scientists keep a detailed lab notebook which contains a record of what they do and what they observe. This may be done in an actual notebook or using a computer. When keeping a lab notebook, you should have one simple goal - someone else should be able to read your notes and perform your experiment exactly the way you did. This gives you a record of what you did in case you forget (you eventually will). It also allows other scientists to repeat your experiment and confirm (or dispute) your results.

GUIDE TO SCIENTIFIC EXPERIMENTS

Step 6 - Analyze Data

The analysis of simple experiments usually has three components:

1. Calculating central tendency (i.e. average) for each experimental group.
2. Calculating variability (i.e. standard deviation) for each experimental group.
3. Examining graphical representations of the data.

Central tendency

Imagine that you gathered the following data:

	Height of plants in dark (cm)	Height of plants in light (cm)
Trial 1	25	12
	27	10
	30	8
	27	15
	23	7

Trial 2	28	19
	32	10
	30	15
	31	12
	25	13
Trial 3	35	12
	27	10
	27	10
	29	12
	26	8

You could use this data to calculate the following averages and standard deviations (we'll explain standard deviations in a moment):

	Plants in dark		Plants in light	
	Avg. ht.	St. dev.	Avg. ht.	St. dev.
Group 1	26.4	2.6	10.4	3.2
Group 2	29.2	2.8	13.8	3.4
Group 3	28.8	3.6	10.4	1.7

Calculating averages makes it much easier to present data in tables and graphs, and to understand the overall trends.

Variability

[Note: This section is for advanced students.]

Central tendency tells you about the middle of a data set, while *variability* tells you about data spread. Variability is often crucial to interpreting an experiment correctly. The easiest measure of variability is the range (the difference between the highest and lowest data points), but the range isn't normally very useful. A more difficult, but much more useful measure of variability is *standard deviation*. Normally, about 68% of data points fall within 1 standard deviation of the average, and 95% fall within 2 standard deviations. For example, if the average is 10 and the standard deviation is 1, then 68% of data points fall between 9 and 11 and 95% of data points fall within 8 and 12.

If an average is 100 and the standard deviation is 15, 95% of data points fall within what two numbers?

> 70 and 130

Scientists use a spreadsheet, calculator, or some other computer software to calculate standard deviations.

You can use the data for Trials 1, 2, and 3 above to figure out how to calculate correct standard deviations using one of these tools.

You may find that your variability is so large that you can't make any firm conclusions. To reduce variability in your experiments, you can do three things:

- Measure more accurately
- Increase the sample size
- Make sure that everything is being held constant except the independent variable

Statistical methods can become quite complicated in larger experiments, and often scientists will ask statisticians for help before they even perform experiments. In particular, it is common to use statistics to decide whether there is a significant difference between experimental groups. For students learning the basics of experimentation, it is sufficient to calculate a basic measure of variability and to keep in mind that more complex statistical methods exist.

Graphing

Creating graphs will help you understand patterns in your data, and will help you explain results to others. Someone else should be able to understand your graph by looking only at the graph and a short title or legend. Here is a checklist for graphing your data:

GUIDE TO SCIENTIFIC EXPERIMENTS

- Choose an appropriate type of graph (bar, line, etc.).
- Label the axes so that the independent variable is on the x-axis (horizontal) and the dependent variable is on the y-axis (vertical). Include units in parentheses.
- Create scales on each axis that accommodate all data points and use consistent intervals.
- Add data points.

Advanced students should also include error bars on graphs to indicate the variability of the data. Error bars are used whenever data points represent averages of multiple measurements. For example, error bars in the graph below extend 1 standard deviation above and below each data point.

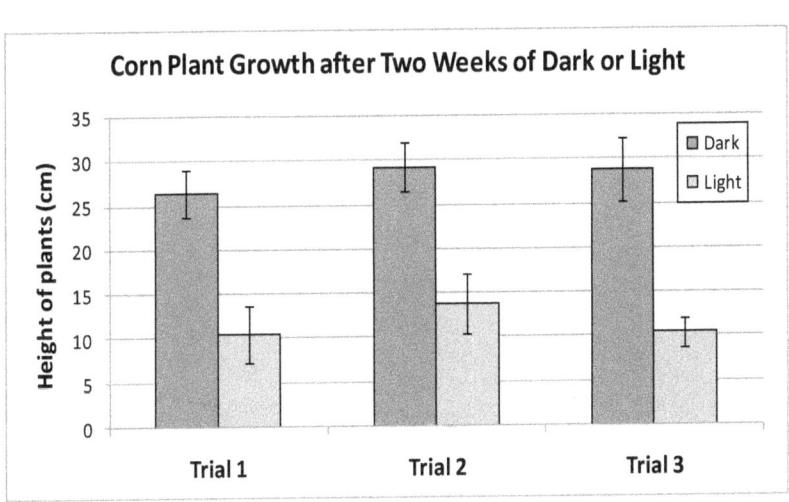

In the first bar in the graph above, the error bars (which represent standard deviation) extend from 23.8 to 29.0. What exactly does this tell you?

> 68% of data points fall within 23.8 and 29.0.

Spreadsheets

If you have access to a spreadsheet program on a computer, then use it. Take the time to learn how to type data into a spreadsheet, use formulas to do calculations (averages, standard deviations, etc.), and create graphs. Find someone to help you if necessary. A little bit of time spent learning how to use a spreadsheet will result in better products, save you tons of time on future projects, and give you a marketable skill for whatever career you choose.

GUIDE TO SCIENTIFIC EXPERIMENTS

Step 7 - Make Conclusions

Scientific conclusions address the original question and hypothesis, and accurately reflect the data collected in the experiment. In the "Ask a Scientific Question" section, you examined the following questions:

- Do plants grow better in light or dark?
- Do plants grow taller in light or dark?
- Do corn plants grow taller in two weeks of constant light or two weeks or constant dark?

If you performed the experiment and found that the corn plants in the dark grew taller, which of the following would be the best conclusion?

- Plants grow better in the dark.
- Plants grow taller in the dark.
- Corn plants grow taller in two weeks of constant dark than they do in two weeks or constant light.

The third option

Specificity is almost always better in science. In the first conclusion, "better" is too vague. The second conclusion could not be justified unless you had tested a large number of different types of plants. The third conclusion is specific enough that it could be more easily defended by the experiment that was performed.

Before making a conclusion, consider the variability of the data and other complicating factors very carefully. If the data are inconclusive, then you should say so. You can then elaborate on how a more conclusive answer could be reached in future experiments.

Questions you should consider when making conclusions include the following:

- Does the data support your original hypothesis?
- How much variability is there in the data?
- Is the data conclusive?
- How does your experiment relate to the larger world?
- What are the limitations of your experiment? How could it be improved?
- What additional questions has the study raised for future research?

Step 8 - Present Results

Scientists usually present results of their experiments in three different ways: written reports, posters, and oral presentations. In every format, the goal is to present accurately, concisely, and in a way that others can appreciate how the experiment relates to the "big picture" of science and the world.

Most commonly, scientific reports have the following sections: 1) Title, 2) Abstract (or Summary), 3) Introduction (or Background), 4) Materials and Methods, 5) Results, 6) Discussion (or Conclusion), and 7) References. Try to match each section below with its description:

1. Title	_____ A. Information about how an experiment was done that would allow someone else to repeat the experiment.
2. Abstract	_____ B. Description of current knowledge about a topic which provides a context for the experiment that has been done.
3. Introduction	_____ C. A short phrase (normally 10 words or less) that gives an overview of an experiment.
4. Materials and Methods	_____ D. Short paragraph summarizing the purpose, methods, results, and conclusions of an experiment.
5. Results	_____ E. Section that contains the authors, titles, publication dates, publishers, page numbers, etc. of all sources used.
6. Discussion	_____ F. Interpretation of data collected in an experiment, including any conclusions that can be drawn.
7. References	_____ G. Description of data collected in an experiment.

> A. Materials and Methods; B. Introduction; C. Title; D. Abstract; E. References; F. Discussion; G. Results

Be sure to read specific instructions about how to format reports for whatever class or competition you participate in, as specifics can vary.

Many science fairs and conferences have "judges" who are supposed to pick the "best" projects. Most good scientists would tell you that picking the best experiments out of a set of very good ones is impossible. Do not worry about winning or not winning any such competitions. The most important thing that comes out of science fairs, class presentations, etc. is that you get constructive criticism on the experiment that you performed. Scientists usually do not think in terms of winning and losing. Instead, the goal is always discovery!

The True Scientific Method is a Cycle

For scientists, one experiment leads to another which leads to another. There is no end to the scientific method. Instead, the scientific method is a cycle. You ask a question and seek the answer, and inevitably the answer raises another set of questions. Human knowledge increases over time as people continue to build on current knowledge by asking more and more questions.

Whatever your experiment, the conclusions will invite another experiment. So flip back to the beginning of this guide and get started again! The excitement of experimentation and thrill of discovery are just beginning!

Guidelines for Achieving Great Experiments

1. Ask a very specific, testable question.
2. Test a control for comparison (a group that does not receive the experimental treatment).
3. Use a sample size large enough to allow firm conclusions.
4. To understand a whole population, obtain a random sample of that population to avoid bias.
5. Replicate each part of the experiment (at least 3 times).
6. Hold all variables constant between trials except the variable being tested.
7. Collect quantitative data whenever possible.
8. Measure using metric units.
9. Gather data carefully and accurately.
10. Be objective and honest.

Scientific Notebook

Scientists keep a detailed lab notebook that contains a record of everything about a project, from the initial brainstorming of ideas to the final presentation of results.

Examples of what may be included in a scientific notebook include the following:

- Ideas for a project
- Notes from meetings with mentors
- Phone numbers, contacts or sources
- List of supplies needed
- Book, article, and website references
- Diagrams and sketches
- Data and daily observations
- Pictures (often taped in) of results
- Calculations
- Graphs and charts
- Interpretations of results

Get in the habit of writing down everything you do. This gives you a record in case you forget what you've done (eventually you will), and it makes it easy to write a final lab report. An accurate notebook also allows other scientists to repeat your experiment and confirm (or dispute) your results.

GUIDE TO SCIENTIFIC EXPERIMENTS

Date:_____ Experiment: _____

Notes:

Date:_____ Experiment: _____

Notes:

GUIDE TO SCIENTIFIC EXPERIMENTS

Date:_____ Experiment: _____

Notes:

THE BLUE HELIX

Date:_____   Experiment: _____

Notes:

GUIDE TO SCIENTIFIC EXPERIMENTS

Date:_____ Experiment: _____

Notes:

THE BLUE HELIX

Date:_____   Experiment: _____

Notes:

GUIDE TO SCIENTIFIC EXPERIMENTS

Date:_____ Experiment: _____

Notes:

THE BLUE HELIX

Date:_____ Experiment: _____

Notes:

GUIDE TO SCIENTIFIC EXPERIMENTS

Date:_____ Experiment:_____

Notes:

THE BLUE HELIX

Date:_____ Experiment: _____

Notes:

GUIDE TO SCIENTIFIC EXPERIMENTS

Date:_____ Experiment: _____

Notes:

THE BLUE HELIX

Date:_____   Experiment: _____

Notes:

GUIDE TO SCIENTIFIC EXPERIMENTS

Date:_____ Experiment: _____

Notes:

THE BLUE HELIX

Date:_____ Experiment: _____

Notes:

GUIDE TO SCIENTIFIC EXPERIMENTS

Date:_____ Experiment: _____

Notes:

THE BLUE HELIX

Date:_____ Experiment: _____

Notes:

GUIDE TO SCIENTIFIC EXPERIMENTS

Date:_____   Experiment: _____

Notes:

THE BLUE HELIX

Date:_____   Experiment: _____

Notes:

GUIDE TO SCIENTIFIC EXPERIMENTS

Date:_____ Experiment:_____

Notes:

THE BLUE HELIX

Date:_____ Experiment: _____

Notes:

GUIDE TO SCIENTIFIC EXPERIMENTS

Date:_____ Experiment: _____

Notes:

THE BLUE HELIX

Date:_____ Experiment: _____

Notes:

GUIDE TO SCIENTIFIC EXPERIMENTS

Date:_____ Experiment: _____

Notes:

THE BLUE HELIX

Date:_____ Experiment: _____

Notes:

GUIDE TO SCIENTIFIC EXPERIMENTS

Date:_____ Experiment: _____

Notes:

THE BLUE HELIX

Date:_____ Experiment: _____

Notes:

GUIDE TO SCIENTIFIC EXPERIMENTS

Date:_____   Experiment: _____

Notes:

THE BLUE HELIX

Date:_____   Experiment: _____

Notes:

GUIDE TO SCIENTIFIC EXPERIMENTS

Date:_____   Experiment: _____

Notes:

THE BLUE HELIX

Date:_____ Experiment: _____

Notes:

GUIDE TO SCIENTIFIC EXPERIMENTS

Date:_____   Experiment: _____

Notes:

THE BLUE HELIX

Date:_____ Experiment: _____

Notes:

GUIDE TO SCIENTIFIC EXPERIMENTS

Date:_____ Experiment: _____

Notes:

Date:_____ Experiment: _____

Notes:

GUIDE TO SCIENTIFIC EXPERIMENTS

Date:_____   Experiment: _____

Notes:

THE BLUE HELIX

Date:_____ Experiment: _____

Notes:

Date:_____   Experiment: _____

Notes:

THE BLUE HELIX

Date:_____ Experiment: _____

Notes:

GUIDE TO SCIENTIFIC EXPERIMENTS

Date:_____ Experiment: _____

Notes:

THE BLUE HELIX

Date:_____ Experiment: _____

Notes:

GUIDE TO SCIENTIFIC EXPERIMENTS

Date:_____   Experiment: _____

Notes:

THE BLUE HELIX

Date:_____ Experiment:_____

Notes:

GUIDE TO SCIENTIFIC EXPERIMENTS

Date:_____ Experiment:_____

Notes:

THE BLUE HELIX

Date:_____ Experiment: _____

Notes:

GUIDE TO SCIENTIFIC EXPERIMENTS

Date:_____ Experiment: _____

Notes:

THE BLUE HELIX

Date:_____ Experiment: _____

Notes:

GUIDE TO SCIENTIFIC EXPERIMENTS

Date:_____   Experiment: _____

Notes:

THE BLUE HELIX

Date:_____ Experiment:_____

Notes:

GUIDE TO SCIENTIFIC EXPERIMENTS

Date:_____ Experiment: _____

Notes:

THE BLUE HELIX

Date:_____ Experiment: _____

Notes:

GUIDE TO SCIENTIFIC EXPERIMENTS

Date:_____ Experiment: _____

Notes:

Lab Report Template

The following template will help you organize a scientific report for a class or science fair.

Title

Abstract

Purpose of study:

Summary of methods used:

Summary of results:

Summary of conclusions:

Introduction

Background:

Question:

Hypothesis:

Materials and Methods

Independent variable:

Dependent variable:

Experimental constants:

Control:

Protocol:

Results

Data collected:

Other observations:

Graph(s):

Discussion

Interpretation of data:

Conclusions:

Reflections on future work:

References

Units of Scientific Measurement

Scientists use metric prefixes and metric units of measurement so that they can be understood around the world. The modernized version of the metric system is called the International System of Measurement, or SI.

Metric Prefixes

Prefix	Meaning
Giga- (G)	$10^9 = 1,000,000,000$
Mega- (M)	$10^6 = 1,000,000$
Kilo- (k)	$10^3 = 1,000$
Hecto- (h)	$10^2 = 100$
Deka- (da)	$10^1 = 10$
Deci- (d)	$10^{-1} = 0.1$
Centi- (c)	$10^{-2} = 0.01$
Milli- (m)	$10^{-3} = 0.001$
Micro- (µ)	$10^{-6} = 0.000001$
Nano- (n)	$10^{-9} = 0.000000001$

Metric Units

Measurement	Commonly used metric unit	Conversion factors
Length	Meter (m)	1 m = 3.2808 ft
Area	Square meter (m^2)	1 m^2 = 10.7639 ft^2 = 1.1960 yd^2
Volume (solids)	Cubic meter (m^3)	1 m^3 = 35.30 ft^3 = 1.3079 yd^3
Volume (liquids, gases)	Liter (L)	1 L = 1.0567 qt = 0.26 gal
Mass	Gram (g)	1 g = 0.03527 oz
Time	Second (s)	1 s = 1/60 min
Speed	Meters/second (m/s)	1 m/s = 2.24 mi/hr
Temperature	Degrees Celsius (C°)	C° = (F° - 32) / 1.8 = K - 273.15
Energy	Joule (J)	1 J = 0.23901 cal = 9.4781x10^{-4} btu
Pressure	Pascal (Pa)	1 Pa = 0.0001450377 lbs/in^2
Amount of a substance	Mole (mol)	1 mol = 6.02214×10^{23} atoms
Light intensity	Candela (cd)	1 cd/m^2 = 0.2919 fl

Glossary

Abstract - Short paragraph at the beginning of a scientific report that summarizes the purpose, methods, results, and conclusions of an experiment.

Central tendency - A measure of the center of a data set. Examples include averages (means), medians, and modes.

Control - A group in an experiment that does not receive the experimental treatment. A control is used for comparison.

Dependent variable - The variable that is measured in an experiment. It is "dependent" on the independent variable, and appears on the y-axis of graphs.

Error bars - Bars above and below a data point in a graph which show the variability. Error bars usually represent standard deviation or standard error.

Experiment - A situation where something is systematically changed and then the effects of the change are observed.

Hypothesis - An educated guess about how an experiment will turn out.

Independent variable - The variable that is changed by the experimenter. It appears on the x-axis of graphs.

Lab notebook - A scientist's written record of everything having to do with an experiment, including experimental design, observations, data collection, ideas, etc.

Mentor - Someone who can give you feedback and advice on your experiment. A mentor should have either experience with scientific experiments or a special knowledge of a subject.

Range - A measure of variability; the difference between the highest and lowest data points.

References - Section at the end of a scientific report that contains the authors, titles, publication dates, publishers, page numbers, etc. of all sources used. References are cited in the text of a scientific report.

Replicate - Repetition of an experiment. Doing replicate experiments confirms results and reduces variability.

Sample size - The number of observations or data points in an experiment.

SI system - The "International System of Measurement." SI is a modernized version of the metric system. Principal SI units include the following: meters (m) for length, kilograms (kg) for mass, seconds (s) for time, amperes (A) for electrical current, moles (mol) for the amount of a substance, candelas (cd) for light intensity, and Kelvin (K) for temperature.

Significant difference - A difference between experimental groups that is large enough to have statistical support. Scientists usually consider a difference significant when there is at least a 95% chance that the groups are different.

Spreadsheet - Computer software that contains a grid for users to enter data, perform calculations, and create tables and graphs.

Standard deviation - The most commonly used measure of variability.

Statistics - The process of summarizing and describing data.

Variable - Any part of an experiment that can be changed. Types of variables include independent variables and dependent variables.

Variability - The amount of spread in a data set. Examples include ranges and standard deviations.

X-axis - The horizontal axis on a graph that shows the independent variable.

Y-axis - The vertical axis on a graph that shows the dependent variable.

www.ingramcontent.com/pod-product-compliance
Lightning Source LLC
Chambersburg PA
CBHW071311040426
42444CB00009B/1970